The Synopsis of Life

Acknowledgments

Cover image used with permission by Rich Murray.

Thanks are given to my family for supporting such endeavors.

Chapter One- The Organism System

A fractal is a pattern of infinite self similar scales. Life is a living fractal spanning throughout the universe. The organic functions of cells, organisms, solar systems, and galaxies serve to highlight this recursion. The actual size of life isn't limited to human kind's stretch, it spills out into the cosmos. This living commotion can be seen in our own solar system. The various biological structures found in space will be eyed and described.

The fractal organization of organic life-forms has created complex systems that have arose to associate in a manner only beneficial for life as a whole. This stability isn't just limited to cells or organisms. The organization of certain scales of life to rely on higher scales of life makes up the fractal repetition template. Without a stable environment organics cease to exist.

All forms of life follow the fractal repetition template being that it is the optimal way life could exist. It's why certain animals evolve certain ways to live in a certain ecosystem. It's essentially the final result of evolution. There are three different types of organic fractal repetition templates. Cells follow the cellular template and organisms follow the organism template. Earth acts like it's own scale of life. The Earth is the environment that both cells and organisms rely on to live, thus following the organism system template.

The scales of life never look the same, but do act the same in function. Their sub components must have similar goals,while their shape or form doesn't have to be similar at all. This is comparable to a computer virus and a biological virus infecting through different forms. If these scales of life mutated to something different in function, they would eventually cease to be stable due to evolution.

The Earth as an organism system is very complex. It is made up of processes that are inter-dependent. This has evolved through time. Humans work together with other organisms to maintain our ability to survive. Earth contains subsystems of organic processes that are required for stability. These subsystems aren't solely the result of living organisms. The earth's mantle and atmosphere maintain some of these means. Without these simple subtle tools, life's framework would unwind in chaos just as entropy suggests. These tools are maintained by strict regulations that preserve order. Our society is one of these tools. Religion tends to instill a peace while local customs ensure it. Media and the mainstream of society is the neural networks of earth's nervous system.

The earth has a vascular system just as we do. Animals and most plants contain a network of vein connections called a vascular system in which material that would normally be depleted in one area is allowed to be sufficient. The need for this system is due to the specialization of different sections of the earth. Some parts of the world specialize in growing raw resources like Africa, while others may specialize in technology, like the United States.

The world is inter-dependent. Without transportation of materials and foods throughout earth, each city would be forced to act as one. This same concept is true for all of life. Animals and plants will die quite quickly if there is a lack of nutrient distribution. Earth has evolved a vascular system of rivers, roads,and great circle routes. The trucks and trains transport goods to where they are needed in a system called supply and demand. Plants and organisms live with supply and demand, but it is an automatic process called diffusion. Whole organisms also account for a decrease in supply with the vascular system.

The heart will contract more if blood is needed.

The earth has a developed lung system. A forest of trees and a sea of green algae serve the role of converting carbon dioxide to oxygen. The lungs of an organism are meant to convert gases into other gases for reaction energy. The stability of this system is vital to survival. The atmosphere of the earth after its initial formation was made of mostly carbon dioxide. Mars is very similar. Once earth started to become more stable and the temperature increased, immature cells emerged. Once they began photosynthesis free oxygen eventually built up. Organisms then began using aerobic metabolism and essentially breathing oxygen. These two components to earth's lung system are what make it stable. The structure of earth's lung system is like two systems breathing into each other simultaneously. The very structure is fractal in image. Without this subsystem, life would eventually run low of CO_2 or O_2 and a cascade of extinctions would result.

The earth's subsystems work in a seemingly long term fashion. The amount of time for processes to finish is significantly longer than the average human lifespan. Without certain subsystems like the lung system, life on earth would still continue living for a short amount of time. These cycles and subsystem processes take hundreds of years to change significantly. The earth's subsystems are thus long term mediators in stability on earth.

Earth also has a complex nervous system. Just like a mammals nervous system, earth has many systems of sight detection. Satellite imaging systems, telescopes, GPS, and weather radars are just a couple of the numerous "neural" systems the earth uses. Eventually these systems will stretch the span of the whole solar system. Humans are the only organisms that directly use them because they are like neurons in the nervous system. The main grid at which this nervous communication system works is called the internet. The birth of the internet will have the earth's infant brain learn to become an adult. Media and mainstream is essentially the brainstorming inner conscious that is driving culture forward. This path of evolution shows that certain ways of life are better for certain time periods and situations. The family tree of different societal ways of life would show the evolutionary aspects our society lives by. Certain ways of life would die out while others would evolve on. Earth's conscious is able to evolve this through “guess and check”. Earths consciousness is human society. Human's are thus the brain of earth. A neurotransmitter is a chemical that can send information from one neuron to another. An idea that someone knows about and that could be spread is called a planetary neurotransmitter. A movement of ideas or planetary neurotransmitters will need to pass a certain threshold just like a nerve impulse. Neurons create and transfer neurotransmitters to other neurons. This exchange allows information to be processed and altered. Neurons also compute and process the information in their dendrites. Just as in neurons and their dendrites, humans create ideas and process given thoughts. Humans acting as neurons for the earth makes us direct parts of the organism system. The earth's nervous system is going to be in function exactly similar to a human's nervous system due to the internet.

Earth also organizes it's nervous-like tissue. Through time humans have developed societies creating social norms dictating certain ways to live. Rules such as a limit on a humans actions on another, are how this organic organization prevents it's humans from killing each other. The equivalent to killing a human in the brain is to induce excitatory neurotransmitters in a process called neuronal excitoxcity. The 10 commandments can be seen in the behavior of neurons. Humans are most stable when they do not steal food or water. This is equivalent to boxing out or vasoconstricting other neurons from blood flow. If all neurons prevented blood flow, the organism would have died out of the evolutionary tree of life. Adultery is similar to certain neurons which are excited greatly by another neuron rather than by its neighboring neuron. The others such as having no beliefs in other gods and not to bear false witness are all dealing with a neurons false signaling or noise-like hallucination synaptic patterns. All religions have a holy day, and this coincides with neurons going to sleep.

Religion is an emotional way to instill stable behavior. Society is an ethical way to do it. Culture is a logical way to do it. We all see these separate things as special distinct aspects of our life but they are all essentially the same. Religions have evolved to bless the group with an increased chance of living. Religion has prevented sexual disease and spread stability. Religion is simply a tool for societal and cultural stability on earth.

The Distribution of neurons to that of other soft tissue cells is similar to the distribution of humans to animals on earth. Humans are misguided to think and focus purely on human kind and ignore supporting organisms. In reality, our importance is nil without supporting animals. Supporting organisms must out number humans by a high factor to support all subsystems like the digestive system, lung system and immune system of the earth. Just like in the brain, humans consume more than they should but still not even half the O2 and nutrients. If we were to replace the muscle bearing animals with lab made food we would be starving many subsystems of the organism system from stimulation. Without these systems, long term stability is hampered. This shows the specialized intelligent purpose we have. Since humans cannot be isolated from other organisms, our function to think is understood. People with the nerve to explore the galaxy in it's entirety would still be part of the organism system serving as nerves in earth's hand feeling around and inside the Galaxy.

Predation is an intricate energy recycling system. We consume organisms throughout the day for our whole life without giving it a second thought. It is what directly ties us to each other. Without predation, over population of oxygen consuming animals would result; leading to instability on the earth. Oxygen levels would diminish and all life would die out. The biggest animals such as elephants, bears, and rhinoceros would fall first due to their higher demand for air. Second to fall would be the rest of the animals. The plants would slowly be starved of water due to their inability to absorb it when there is a deficient oxygen environment. The last organisms would be sea faring. They would be the ones to "save" life on earth because atmospheric dissolution is a rather delayed process. Therefor they would still have access to valuable oxygen stores. At least 10 of these planetary instability events have occurred in history. Without great extinctions occurring, earth would have never evolved to have organisms that regulate oxygen, carbon dioxide, or water. Earth would have never evolved to have organisms that regulate the oceans salt. Earth would have never evolved to have organisms that regulate nitrogen. Earth would have never evolved to have organisms that regulate the earth's surface temperature. Earth even would have never limited the aggressiveness of the species to prevent over population. Earth wouldn't have lived on to be it's own organism system. Earth would have just kept having a great extinction event after another.

Earth's reproductive system is very complex while

the solar systems replication method is more simple in terms of understanding. An organism system will reproduce separately from the replication of a solar system as will be explained in the next chapter. The possible sexual reproduction between two different planets resulting in a third planet is different than the asexual reproduction of just earth. Both methods of reproduction are possible for earth. Earth's reproduction and conception require both terrain-forming and geoengineering. These processes change the planet to suit their needs. Essentially the processes that must be set up are the subsystems that the organism system needs to survive in the long term. This new planet must have certain features the original planet finds "attractive" or suitable. When a group of humans is sent to populate the planet, these "sperm" bring information and begin to change the nature of the planet. A stable respiratory system must be setup. A stable nervous system must be setup. A stable vascular system must be setup. Eventually when all subsystems develop, a stable organism system forms. This brand new baby has the same main DNA or information but just as in organisms mutations can arise. Geological factors, chemical factors, or societal factors cause mutation. They could be cancerous and dangerous or of course beneficial.

Everything on earth supports the stability of the organism system. Organisms are essentially tame to the earth. Earth has evolved to control the different specialization each organism must go through. If earth was not living you would expect life to be universally competitive similar to the initial conditions of life on earth. The result of controlling animals behaviors such as predation is to instill a fluid stability onto earth. If all animals were able to eat any plant, specialization would not occur. All the animals would be adapted to consuming any plant. Since changes in the animal wouldn't benefit it, only one similar type of animal would emerge. Each animal species has a selective diet of predation. Each animal has evolved around that diet. Cattle have adapted to eat grass yet many other animals are unable. Some plants are poisonous to certain animals yet aren't poisonous with other animals. Stability systems such as the oxygen cycle,water cycle,nitrogen cycle, and carbon cycle would never develop when there is a lack of specialization because animals would eventually counter balance everything. Certain animals that would be positive to the environment would be just as prevalent as ones that are negative. Apex predators have become able to benefit from phytonutrients in the plants. These same chemicals are toxic to organisms on the lower levels. This differentiation allows subsystems and stability to form. If they are not beneficial, a negative feedback will eventually result. Thus the immune system will clear and cough up the problem.

The Immune system of the earth is ironically the virus' and toxins that are on earth. Unstable animals are more susceptible to virus'. This is equivalent to an unhealthy cell that the immune system needs to remove. If overpopulation of an organism in one area like cancer occurs, the immune system will try to kill them by quickly spreading viral outbreaks. The immune system on earth is the same as in an organism, it kills off things that would make the total organism unstable. Earth's immune system can also store information about current major infestations like a macrophage does to lymph nodes by being passed down into the genes of an organism by benefiting the health of it's embryo. The majority of our DNA is viral DNA. An antibody is a chemical that is used to identify and neutralize a target antigen. An antigen is a chemical that will attach to an antibody. Antibodies in the Earth's immune system are the virus'. An Antigen in the earth's immune system is an organism's susceptible immune system. Toxins are the virus' method of dealing with the problem. Poisonous plants are also a small part of the immune system by killing off unhealthy animals that are scavenging to eat. They kill animals that aren't behaving in a normal way. This is similar to an animal that is mutated or cancerous. This prevents instability. Inflammation in earth's immune system is essentially habitat degradation. Just as in an organism, some inflammation is necessary, but too much is dangerous. The cause could be from a meteor,fires,volcanism, desertification,or deforestation.

Earth's digestive system works in the same way as an organisms does. Decomposers first chew away the main aspects of the dead organisms just as in the first pass in the stomach. The fungus and bacteria then break down the trace remnants. After these first two pass exchanges, very little is left to pass through the last process. Pressure decomposition is what makes crude oil and eventually converts the bones in it back into a type of rock. The breaking down of organisms recycles the nutrients and delivers the nutrients to the earth. Just like with the earth's lungs being to lungs inside each other, the digestive system is essentially like two stomachs inside each other.

Earth's Skeletal system is a framework of buildings, habitats, infrastructure, mountain ranges, caves,and bridges. Essentially anything in the mantle of the earth could be part of the skeletal system. For humans, earth's skeletal system is mainly what we have constructed. Like the skull of earth, our cities house our workplace and residence. The strength of these cities are becoming greater and greater through the maturation of the organism system. For animals, their habitats rely on primary parts of earth. They don't specialize for extra protection because they don't need it. Humans need it because they are weak and specialized in intelligence.

The similarities between organisms and the earth show the basis of the organism system template. Out of the three templates, it has the most restricted role in stability. Compared to organisms, Earth is in a relatively fixed place. The relation of the organism system to a cell and an organism is limited purely to function, as it wouldn't make sense if each scale of life had the same aesthetics. An organism system is like a cell and organism combined. Essentially you could point to any part of your body and it can be related to a part of the earth, even if they don't look similar.

Chapter Two- The Solar Cell

Earth's subsystems are necessary for it's stability. The similarities between an organism's living functions and Earth's stability functions entangles the idea that Earth is living. The huge system that the earth is a part of is called the solar system. The solar system is able to maintain a strict and stable formation. Just like the earth and the organisms on earth, the solar system is made up of subsystems that prevent instability. Space is a dangerous place when detached. The darkness creeps with instabilities.

The sun isn't the only celestial body in the solar system that the earth counts on to survive. Every part of the solar system is needed for life on earth to continue, even the moons. The earth is to the solar system as an organelle is to a cell. An organelle is a subsystem that is used to perform specific actions. The solar system and it's subsystems act together exactly as a cell does.

The actual properties of the solar system are largely unknown. Few moons and even fewer asteroids have been studied. Each of these could have novel processes with significant systematic effects. The atomic distribution gradient allows different planets and moons to have different compounds. Natural processes on these bodies are very common. Volcanic out gassing is one very common reaction. Each of the few things we do know about the solar system seem to be novel discoveries.

The Solar System cell has been lit up with recent research tying it together with a chemical, atomic, electrical, magnetic, and biological framework. The mainstream view of the solar system is deficient. The framework of the solar system is exactly similar in function to that of a cell. The solar system isn't a dangerous random place, it is quite stable. With knowledge the void is no longer dark.

Every biological cell in history has had a membrane serving as a wall that isolated it from the outside environment. The solar cell also has a membrane. Cells use this membrane for a variety of functions. The main use being a protective barrier. The solar cell relies on this membrane to protect itself from violent ionized particle bursts and poisonous ion chemical streams. During supernovas and other galactic phenomena, gamma rays are released at nearly the speed of light. Cells on earth need a solar membrane to avoid these threats. Without this membrane, life wouldn't exist now nor would it ever. This is because the energy of the outburst would have stripped any planets atmosphere as well as have induced photodissociation of any complex particles required for life. After an event like this, complex particles would become simple. The earth would lose nearly all of it's living organization. The cells would be entirely torn apart. Every animal and plant would die immediately. A similar concept that humans use when sterilizing things is called ultraviolet germicidal irradiation. This form is weak and ineffective compared to natural gamma ray bursts.

A traditional cell is the original cell structure biology has studied. The term traditional cell refers to the common organization that all cells on earth share. A traditional cell is very similar to that of the last universal ancestor except that it is more evolved. These commonalities such as relying on ATP for energy are significant because they are only prevalent because they are optimal. A traditional cell is thus the template for life at the cellular level. This organization isn't limited to the relative particle world we normally see cells in. The cell template can prevail in some circumstances such as physical systems and information systems.

The biological structure of the solar membrane has similar components to it's structure on earth. Traditional cells have a lipid bi-layer membrane that is extremely thin. It is composed of two molecules with a bubble head facing outwards. These two molecules are very fluid yet sturdy. The solar membrane in astrophysics is called the heliopause. The sun's magnetic field combined with the solar wind ions create one component. The other Component is from the incoming interstellar wind. These two components form two polar regions with a neutral inner region. These two polar components of the solar membrane create bubble shaped fields of ions as shown by space probe research. It has the same components as a traditional cell's membrane. The magnetic solar wind from the sun is pushing against the magnetic local inter stellar wind just as the outside of a traditional cell is always pushing inside.

These ions form a membrane due to their chemical properties. They are ions because the total number of electrons is not equal to the number of protons, giving each particle a positive or negative electrical charge. The solar membrane blocks most ions from passing through. Examples of ions which are in the solar system are oxygen, hydrogen, and carbon. The solar membrane is thus an electrical barrier from highly energetic destructive cosmic rays. The solar membrane is selective with what it blocks out and willingly allows diffusion of low energy neutral atoms like that common on earth. Diffusion is hallmark to cell membranes.

The solar membrane is located around 100 AU from the sun. One astronomical Unit, or AU, is roughly equivalent to the distance from earth to the sun. Neptune is about 30 AU from the sun. Past Neptune orbits hundreds of thousands of rocky asteroids. These objects orbit the sun solely. The Kuiper Belt as well as other groups of asteroids are between 40 and 100 AU. Outside of this, or outside of the solar membrane, lies the oort cloud. The oort cloud is made of billions of comets stretching over a light year or 50,000 AU from the sun. In traditional cells, fluid surrounds the cell to allow adhesion and diffusion of nutrients. Essentially, the fluid around the solar cell is the oort cloud.

While the solar membrane allows neutral particles to readily diffuse across the membrane, ions must be imported by certain mechanisms. Without these mechanisms, there would be a deficit. In traditional cells, one of the most common ways to import chemicals is called protein transport. Protein transport is a group of mechanisms for importing different compounds. The protein transport region of the solar membrane is both the inner Kuiper belts objects as well as the oort cloud objects. The objects which orbit transitions between the different sides of the solar membrane are the actual active transport proteins for the solar cell.

They orbit from colder to warmer regions allowing water to unfreeze causing significant amounts of ice weathering on the rock releasing compounds like water, ammonia, silicates, and hydrocarbons. Asteroids facing the sun at the distance of earth's orbit have a surface temperature of about 100 degrees Celsius, while the dark side has a temperature of about -76 degrees Celsius. This variation is enough to boil and freeze water with each rotation. The interplanetary dust cloud is a widespread diffusion of cosmic dust in the solar cell. The interplanetary dust cloud is evidence of this ice weathering. The significance of these imported compounds such as hydrocarbons and water will be outlined later in the book.

The solar cell processes are relatively slower than in traditional cells. This relative difference in speed is due to the difference in scale. A higher relative scale will seem slower than a lower relative scale. For example, the amount of time it takes for things to be recycled in the solar cell may seem slow since in a traditional cell this process is seemingly instantaneous. Many processes that occur in the solar cell take so long, they don't even finish in a humans lifespan. This fact is one reason that this theory has gone unrecognized.

Eris is a dwarf planet around the size of Pluto. It is a scattered disk object that orbits from 40 Au inside the membrane to 100 Au in the middle of the solar membrane. This area is a dense and perfect region to pick up atoms due to their lack of excitation as evident by voyager 1. Sedna is a large dwarf planet sized object with a long orbit spanning from 80 AU inside the membrane to over 900 AU outside of the membrane. It's relatively long orbit takes over 11,400 years to complete, but this orbit doesn't hinder it's ability to transfer particles from outside the ion membrane to inside. Through life's existence on earth it has made 311,403 trips between the sides of the potential. From 80AU another asteroid or dwarf planet will pick up any ions released from ice weathering or radiation belt emission.

The significance of protein transports in the solar cell can only be lightly explained until other mechanisms are shown. Essentially, important molecules from other solar cells are propelled in a flux called the interstellar medium. These ions need a carrier to push through the solar membrane. These molecules would be deadly and unstable if they were allowed to just impact earth. They could also be toxic if large amounts of toxic gases were allowed to just diffuse into the solar cell onto earth. Many of these different toxic compounds, like hydrogen peroxide, have been observed in space. Asteroids and dwarf planets are examples of carriers for protein transport in the solar cell. Asteroids and dwarf planets absorb and transport bigger molecules as well as ions into the solar cell by passing through the solar membrane. This allows diffusion to be a slowed process allowing acclimation.

Signal transduction is a process at which outside ions can cause an intracellular effect or response. Signal transduction occurs when a solar receptor is activated. The solar receptor is activated by the binding of incoming ions to a dense cloud of protons trapped in the non-motile region of the solar membrane. If the electron doesn't bind to a proton then it is blocked from entering the cell. The solar receptor is shaped like a ribbon around the solar system. Ultra high energy electrons from outside the solar system have affinities for certain solar receptor effects. The incoming ion's energy changes the intracellular effect. A lower energy electron that activates the receptor will travel through the solar membrane and reach only to the gas giants until their speed is slowed down enough to be picked up by either one of the gas giants. Thus they must have a greater "escape velocity" of these planets if they were to reach earth. This essentially is like binding affinities in drug pharmacology. An ultra high energy electron that activates the solar receptor by binding with protons in the non motile region will have enough energy and velocity to strike earth with greater force than any particle accelerator on earth. These collisions occur all the time and have been studied diligently. These ultra energetic neutral particles explode in our atmosphere with great force. Some of these particles have the same energy as a baseball being thrown. The effects of these collisions cause the creation of exotic particles with specific functions. Many particles are formed like neutrinos which are unlikely to interact with ordinary matter. The other array of particles are currently being studied. This new research will lead to expansions in the theory of the solar receptor complex.

IBEX a spacecraft laboratory has been build for the sole purpose of mapping these neutral high energy particles. This experiment discovered a ribbon of activation, or the solar receptor. The actual shape resembles a G protein, like that in a cell. Without this shape, particles wouldn't hit our flat solar cell.

So a certain frequency neutral particle will induce different solar signaling pathways. The different frequencies that particles exhibit are thus the different drugs. The exact pathways are often difficult to particulate due to their lack of absoluteness and pure tendencies.

There are two main types of transduction in a traditional cell, protein modification and phosphate cascade. In the solar cell, protein modification equates to changes in a planet or subsystem and phosphate cascade equates to hydrogen energy cascade. The energy cascade pathway will be explained later during the mitochondria section. The protein modification pathway is not related to protein biosysthesis as there are no RNA being decoded. The pathway mainly deals with activating enzymes for functionality.

Protein modification in a traditional cell is an understood and vital part to our bio-chemistry. Epinephrine is the adrenaline drug, or the fight or flight initiater. It has a few major effects on neuron cells that are distinct from other cells, but epinephrine has effects on nearly every type of cell.

Epinephrine activates multiple major pathways as most drugs do. Epinephrine acts on every type of cell. Epinephrine is able to release glucose or densely stored energy in an indirect fashion. The pathway starts when a group of cells called the adrenal glands expel epinephrine into the blood medium. It binds to G proteins on the cell membrane and the g protein releases a sub unit part of the molecule. This moves to the protein adenylyl cyclase converting ATP to cAMP. This cAMP molecule binds to regulatory parts and releases catalytic portions. ATP then binds and allows the phosphorylase enzyme to release glucose. This traditional cell model is the template for the solar system's receptor signaling.

In the solar cell this exact pathway occurs all the time. Ultra high energy particles with specific frequencies having increased tendencies to land in Saturn's magnetosphere are released from dense galactic centers. The galactic center is equivalent to the traditional cells adrenal glands. The high energy ions bind to the ribbon receptor protons or G protein equivalent and release a subunit or a small part of the ribbon receptor. This new particle is neutral with similar properties as the original high energy ion that was released from the source. This subunit travels towards Saturn being pushed closer every second of it's trip due to its high gravity. Saturn's mass is 95 times earth's mass. Orbiting Saturn is Titan, the 2nd biggest moon in solar system. All moons in the solar cell are enzymes. They act as enzymes because if they weren't there, the solar cell would rely on these particles to just collide randomly and react. An Enzymes job is to speed up this reaction. Gas giants are different because their lack of core makes them able to only catalyze certain reactions.

The next step in the signaling pathway occurs when the neutral particle explodes into either Saturn's rings or Saturn's upper atmosphere which both are almost entirely made of water. Studies have shown that this rain of ionized water particles from the rings are flooding 30 – 43 percent of Saturn's upper atmosphere. This is equivalent to releasing the catalytic portions in the traditional cell's signaling pathway. These water ions now are able to stay in the magnetosphere and fly around the planet at nearly the speed of light. Titan in the orbit of Saturn's magnetosphere then absorbs these ions due to it's own personal magnetosphere. The H2O ions then form H20 neutral particles when combining into the atmosphere. The Titan enzyme then uses water to activate the processes during the degradation of it's methane lakes.

We most likely don't see the enzyme of this pathway in its prime because we shouldn't assume energy reserves are currently being completely used up. If we looked into our cells right now, this pathway wouldn't be in much use compared to during an active flight or fight response. This pathway is not a specific pathway since it can cause other responses other than a release of glucose. This coincides with many cellular receptors in traditional cells. We could discover every potential solar cell pathway because a lot if not all are universal pathways required for the optimal function of the organism.

Titan, Saturn's moon, is a phosphorlase enzyme. It could have a thriving bacteria population, but we can't be sure because we have no way to feel if epinephrine is abundant. Titan will never have humanoids because its job isn't to make proteins or machines. Now if this job changes, it's possible. When the sun swallows earth due to expansion there will be no other natural ribosomes other than titan.

In a traditional cell, there is an internal fluid of nutrients called the cytoplasm. Inside the solar cell is a rich fluid of plasma ions
emitted from the sun. This plasm or plasma is thick with a variety of molecules, atoms, and ions. This plasma has water vapor ions, carbon dioxide ions, sulfur, oxygen,

carbon, nitrogen, phosphate and others with hydrogen being the most prominent. These different particles are able allow a wide range of interactions and mechanisms to be common place. Each particle has it's own stable isotopes which also change it's properties. The cytoplasm serves as a gradient displaying the different boiling points of molecules and atoms. Heavy atoms such as oxygen and water will be seen throughout the galaxy on planets situated just like our own. This is because these heavy elements would escape the atmosphere on any closer or smaller planets. Astrophysicists call this distance in the gradient the habitable zone. This is where water stays liquid. If a planet were any closer the water and oxygen would be lost out of the atmosphere as a gas.

Inner planets are rarely gas giants because for one, during the solar systems formation heavier objects tended to stay toward the center while lighter objects tended to be expelled farther away. The second reason is that lighter gases can reach escape velocity from a planet extremely easily compared to heavier gases. This gradient thus is universal as other organic structures in the universe. This gradient makes each different planet specialized in the reactions they need to perform.

A traditional cell structure is formed by the cytoskeleton. It supports the cell like a skeleton would. Without the cytoskelton, the cell would behave like jelly. Eukaryotic Cells have microtubules that maintain cell structure by resisting compression. Microtubules are also the tracks for intracellular transport. These microtubule tracks work the same way as train tracks do in cities.

In the solar cell, the cytoskeleton is formed by the combination of magnetic field connections and solar ions. The magnetic connection the solar system is made of is 100 times stronger than just a magnetic field due to the addition of the solar wind. The solar cell would be miniature and tight in shape without these two components. Most planets and some moons have their own magnetic field called a magnetosphere. These magnetospheres help keep the shape of the circular solar cell.

Magnetospheres also interact with the host planet. The radiation belt circling a planet gives ions energy and speeds them up. Each planet shows auroras when these ions precipitate down to the atmosphere. These magnetospheres also cause particle release upstream towards the sun. In the case of Jupiter's magnetosphere, it's upstream electron bursts reach all the way to earth and can be detected with any radio telescope.

Each of these magnetospheres consists of a magnetotail. The magnetotail will reach an extremely far distance behind the planet in the opposite direction of the sun. The magnetotail pulls ions in towards the source planet from, what is in some cases, the distance between planets. For example Jupiter's magnetotail reaches Saturn's orbit. This allows Jupiter to absorb ions from Saturn's magnetosphere, as well as, from Saturn's trojans and particle backdraft. Most planets have asteroids called trojans that orbit a similar orbit as the planet, but trail behind them allowing them to prevent a loss of particles.

In the developed cell there is a component called the microtubule organizing center. For immature cells, the nucleus serves as the microtubule organizing center. The sun is this exact nucleus organizing center. The source of all magnetospheres is from the sun and all other magnetospheres are dependent on the sun. This organelle also works during cell mitosis or cell splitting. In the solar cell it does the exact same thing since when a super nova commences the sun's core or source of magnetosphere changes shape or collapses.

The Sun is a structure that acts as the solar cell's nucleus. In traditional cells, the nucleus core creates DNA and transfers it into mRNA. This messenger RNA travels into the cytoplasm. From the cytoplasm, ribosomes convert this RNA into amino acids and then into complex protein machinery. RNA is thus the building blocks of proteins in cells. Proteins constitute only 20% of a cells weight, but are vital.

The sun's core creates heavy elements like oxygen, carbon and nitrogen in a process called stellar nucleosythesis. The sun eventually releases these atoms as solar wind. Solar wind consists of protons, electrons, oxygen, nitrogen, carbon, and extremely minute amounts of 67 other elements. The amounts in any given sample are pretty small, but because the size of the sun is 99.85% the mass of the solar system. This small percentage becomes significant when amplified and concentrated. This higher concentration occurs on earth because it's distance and size from the sun allows it to support heavy atoms, where as mercury is too close to the sun to support any significant amount of gases at all. In fact mercury and Venus the first two planets are made of 95% CO_2, one of the heaviest natural gases. This concentration mechanism will be outlined later in the book.

The solar cell has a DNA genome comprised of four compounds. The solar cell DNA is Oxygen, hydrogen, carbon, and nitrogen. These four compounds are similar to Adenine, Thymine,Guanine,and cytosine, respectively. The exact concentrations of these four particles in the sun are the information of the solar cell or DNA. The specific number of atoms constitutes whether certain functions occur in the cell, and in which way. If the sun was made of different amounts of elements, chances are, the cell would change. These four have the most impact ,but changes in other chemicals can cause gene expression changes. A cell with less iron equals a less potent magnetosphere. A cell with less oxygen means a dramatically different or deficient energy source.

Just like in cells, the solar cell formed much of it's protein structures prior to finishing replication. The main structures of stability have already been constructed. The planets formed way before the solar cell stabilized. The asteroid belt and the magnetospheres were all finished before any planet was stable. The mRNA thus at our time is a supporter for the machinery. The mRNA is the solar wind, and earth is one of the ribosomes. Earth thus uses the sun's atomic percentage distribution information to create certain proteins and machinery. The main use of this information is to create a stable environment that can make machinery. Solar mRNA Gives plants energy and warms the earth above freezing. The earth relies on the solar wind because it has nothing else to use. A similar concept is that in traditional cells, a ribosome translates the organisms DNA because there is no other source when isolated. Humans make machinery because it's the optimal way to live. Evolution will nearly always occur when certain ingredients are combined with given time. If evolution creates something unstable, it will eventually die off.

The Earth is a ribosome in the solar cell. A ribosome uses DNA information from the solar nucleus to make proteins. They are essentially the decoding organelles. Ribosomes have been found to be in two forms, bound and free. A bound ribosome is equivalent to a planet in orbit. A free ribosome is equivalent to a rogue planet. Rogue planets have been found in our galaxy, but not in our solar cell.

Cell ribosomes need ATP energy and mRNA to transcribe their proteins. This is because energy is needed to catalyze the reaction. Earth as a whole uses energy from water and the solar wind to create proteins in a time delayed reaction to further our stability. The tools and machines that are made as a result of the evolution on earth are the exported proteins as seen in cells. Factories,labs,farms,mines,schools,offices,and hospitals are examples of the decoding complexes that improve the stability of the solar cell. All of the machines or tools we make are equivalent to the proteins made by traditional cells in their ribosome. Our factories and corporations create vital structures and machinery that will soon serve as part of our very solar cell. Examples of these machines could include, data information satellites, space telescopes, space detectors, space stations, hydrogen energy harvesting, mining facilities, and membrane receptor facilities. There will be countless more inventions or tools that will make our presence in space more prominent and organic.

Earth has functioned as a ribosome ever since the start of life on it. Life is essentially the decoding mechanism that a ribosome is known for. They decode what is optimal for life to exist on a planet with a specific concentration of different particles and molecules. This decoding is essentially evolution.

The mRNA reaching earth in the form of solar wind is only the current gene expression. The solar wind MRNA isn't static because the exact concentration of atoms in the sun aren't present in the solar wind due to fluctuations in the concentration. Solar wind acts as the expressed genes in the solar cell. The expressed genes can be influenced by certain factors. Typically in traditional cells when a cell divides, two cells result with different activated genes for making certain proteins, even if they are part of the same organism. The organism's genome stays the same while the active genes are different. Sometimes a cell can replicate and retain the same activated genes creating a self replicating loop. This living epigenetic pattern maintains it's identity and passes it on. In the solar cell, this exact phenomenon is what a serene race replicating would be like. For example, we would behave this exact way if we were to replicate the solar cell into an additional cell in the absence of foreign cultural convergence. Foreign cultural influences are what would change the epigenetic pattern of our solar cell.

Organic chemistry has already outlined the effect of certain small chemical combinations on their biological and chemical properties. All chemical combinations that contain a certain functional group will tend to have similar properties. These functional groups are very simple and consist of the exact atoms as found in solar wind. A concentration of 1 carbon to 4 hydrogen can create methane with an alkane functional group. A concentration of 1 oxygen to one hydrogen can create ethanol and wood natural alcohol with a hydroxl group. Both of these compounds are found naturally throughout the solar cell and on earth organisms create these compounds as waste. A concentration of 1 nitrogen to 2 hydrogen creating the amide functional group as well as a carboxlic acid functional group combined make amino acids. Carboxlic acid functional groups are the most common acids. Functional groups act as the solar cell equivalent to amino acids. This is because functional groups are essentially the building blocks of life on earth. In a traditional cell, fatty acids and proteins are dependent on functional groups to react. Earth serves as a ribosome in the solar cell and a ribosome is a protein in a traditional cell so the relation equates.

The earth evolved to become a ribosome due to it's placement. The effect of the solar wind on a planet decreases with distance. If earth was much closer to the sun than it is now the solar wind would strip water, oxygen, and other vital chemicals from it's atmosphere and surface. A region, called the habitable zone, is where a ribosome must be to operate correctly. This is why earth evolved to be the ribosome, it is in the gold spot for all it's functions. This placement isn't novel, as Jupiter and most other planets should be where they are. Studies have shown that it is common for a Jupiter sized object to be located right next to an asteroid belt just as it is in our solar cell. There seems to be very little about our solar system that is truly novel. A space telescope has shown that one in five stars similar to our sun have a planet orbiting in the habitable zone.

Traditional cells use chemical energy to speed up certain reactions to maintain a chemical balance. ATP has so much stored energy due to it's two phosphate bonds. Ribosomes need ATP to function. The solar cell uses hydrogen to speed up reactions because hydrogen is such a reactive light weight atom. It's temperature increases much faster than heavier ones. It also is more reactive than other atoms. The thermal temperature of a planet would cause hydrogen to escape if it were not bound to another chemical. This is why Jupiter is the closest planet to have hydrogen gas. Closer planets like earth no longer have free hydrogen. Bound hydrogen is a molecule of free hydrogen combined with another chemical changing it's properties. Hydrocarbons like methane or gasoline are examples of bound hydrogen energy sources. Water is the simplest form of bound hydrogen. Water in the solar cell is equivalent to ATP in the traditional cell. Water is thus the main hydrogen energy source in the solar cell.

In traditional cells, ATP is a compound used in a large number of enzyme reactions. Water is used by organisms as a substrate and medium for chemical reactions. Lifeforms cannot live without water. Some can, however, live without air in certain circumstances. DNA and RNA must be in water for the polymerization reaction in replication to occur. Still, organisms can stay dormant without any additional water. The significance of water on the solar cell level is equivalent to ATP on a cellular level.

Water has become the main source of stored energy because it is the easiest to form and transport to earth due to properties caused by the distribution gradient. Comets traveling towards the sun won't release it until in the vicinity of the earth due to it being in the habitable zone. Active transport in cells also require ATP to function. Chemicals that have trouble being transported upstream to earth will never be developed as useful. Compared to other similar chemicals with similar weight, water has a high boiling point. Water is also able to serve as a carrier for organic acids and alcohols. Water has developed to be the energy carrier or ATP-like substance.

In a cell, mitochondria serve as energy producing components for the cell. This organelle is enclosed by a membrane and appears to function like a limited cell because it breaks down some chemicals to produce ATP. Glucose is split apart in the cell and made into ATP a couple of ways. Glucose thus acts as an energy storage chemical. Without energy storage molecules, the cell would be restricted to the point of death.

In the solar cell, Jupiter and the other gas giants serve as energy producing planets. They act like giant chemical power plants. These giants produce energy compounds due to their heat, pressure, and composition. Water and other hydrogen carriers have been found on the gas giants. The gas giants have the biggest magnetosphere's besides the sun. The magnetosphere is equivalent to a traditional cells membrane. A traditional cells mitochondria has a membrane as well. The gas giant planets and their satellites have more than 10 Times as much water as earth, but uncertainty could even make this number as high as 20 times. Jupiter contains less oxygen than the sun, so water production is predicted to be the source of this. This could mean a lot of Jupiter's H20 has been expelled due to gas currents reaching lower altitudes. When the water vapor is made in the high pressures near Jupiter's core, the gas currents would accelerate to escape velocity. The gas would then fall into Jupiter's powerful magnetosphere and then one of Jupiter's 67 moons. Jupiter's magnetosphere has been found to be rich in H20 ions. The ions continuously circle Jupiter at a high speed until they crash onto a moon or are ejected. Jupiter has been shown to emit particles upstream this way.

Ten gas giant moons have significant oceans. Europa has been predicted to have 3 to 4 times the amount of water as earth. Two other moons Callisto and Ganymede of Jupiter appear to have similar amounts as Europa. Rhea, Titania, Oberon,and Triton have also been found to have significant potential oceans or vapor. Many of these pools of water are thought to be heated by the internal heat of the moons core. This allows them to stay just above freezing.

Traditional cells create ATP in a couple of ways. Glycolysis makes ATP by combining certain chemicals to form ATP. This process occurs in the cytoplasm. There are two main sub processes of glycolysis. When there is sufficient oxygen, a process called aerobic glycolysis will use oxygen and glucose to form ATP. If there isn't enough oxygen then another electron acceptor is used to form ATP. In the solar cell, glycolysis is equivalent to the creation of water from methane. Methane thus serves as glucose. Glycolysis can occur anywhere in the solar cell besides the planets. These processes occur on moons such as titan.

Glycolysis isn't the only process found to produce water. The ATP sythase enzyme is the main method of ATP creation in traditional cells. It takes a used ADP molecule that is missing a phosphate group and converts it into a charged ATP molecule. This must occur in the mitochondria. In the solar cell, this is equivalent to the gas giants. Hydrogen and oxygen is made into water by one reaction. Hydrogen is the phosphate while oxygen is the adenine molecule. The gas giants core acts like the ATP sythase enzyme because the core is where water is formed. This recycling mechanism produces most of the water in our solar cell.

Ammonia and methane have been found throughout the solar cell. Ammonia or NH3 and methane or CH4 donate their hydrogen atoms to produce H2O or water. Hydrogen gas is too light to stay on any body besides Jupiter, Saturn, Uranus, Neptune due to atmospheric escape. Ammonia and methane thus are necessary components in the creation of water in places other than the gas giants.

Ammonia and methane are able to be liquids on nearly all planets and moons outside of earth, and thus are able to react with oxygen. When combusted, methane and ammonia form water. This combustion can occur from volcanoes,meteor strikes ,or lightning.

In a traditional cell, most ATP is recycled with very little relative production being due to electron break down. Water is recycled with the water cycle very efficiently on earth with very little electrical break down or planetary escape.

Everything is recycled in traditional cells. This facet is what helps establish stability. In the solar cell, atoms are recycled readily. The recycling of energy is a common motif in life.

Chapter Three- Maturation of Our Solar Cell

As humans, we can understand what we depend on to survive, but are we able to further help our solar cell? We cannot help the solar cell with tedious tasks like sweeping the floor, but we can in fact evolve or finish maturing our whole cell. We can create astronomical structures that will be like the pyramids of space. We will design elaborate planetary machinery. We are already progressing in the right direction. The cell needs this additional support because not everything has translated from the cell to the solar cell. Proteins can make a weakness' more stable or even change the complete mechanics. An example of this major change would be similar to when cells switched from prokaryotic to eukaryotic means. The DNA structure changed, the tubule structure changed, the nucleus as a whole changed. The whole interaction with other cells has also changed, for example the cell to cell receptor complex.

Hydrocarbons have been found to exist throughout the interstellar medium due to a feature seen on all astronomical objects in our galaxy. The interstellar medium is the space between our star and others, including the oort cloud. Fullerenes and hydrocarbons are so common that certain wavelengths are reduced due to the compounds absorbing this spectrum of light. This fact can help us see the importance our outside has to us. This supply of nutrients acts in a similar function as our blood does. The hydrogen is like the plasma portion and the fullerene is like the blood cell portion. They are similar to blood cells because they can carry small molecules inside their structure, protecting them from breaking down and assisting in transport. These Endohedral fullerenes are natural, easy to make, and can contain water, oxygen, and nitrogen as well as other particles in clusters. They can import ions into our solar cell due to their neutrality. Chemical reactions can cause the fullerene to open up and release the compounds. Similar reactions can also re-close them.

With a blood supply and transport system comes with it other cells and a blood supply source. The multi cellular nature of our organism is apparent. Each star is a different solar cell. These cells are not too relatively similar to ours on average due to their function. Most cells are specialized towards performing one thing. The wide variety of cells make these cells inter-dependent.

The isolation of one solar organism is never perfect. In fact there are more bacteria cells living, probably in symbiosis, than there are traditional cells. The way a neuron cell looks is quite different than a skin cell and that is quite different than a muscle cell,bone cell,liver cell, lung cell and basically any other cell.

There is no such thing as a normal cell, at least in a multi-cellular organism. You cannot point to one cell in one spot and expect an exact specimen. That's what makes us living. Uniformity isn’t a characteristic of any scale of life, only the base goal or operating system. Uniformity thus isn’t the base goal of an organism, stability is.

In our solar cell, we see how our existence has been put on the shoulders of other cells and held up. Human's need to put the team on their back. Our combined effort isn't uniform, but our stability goals are clear.

Aliens understand this concept and do not look for the "life" of the ribosome, but they do look for the maturation of a cell. They do this by communicating with one specific cell to cell receptor. Communication could take as little as two years to reach. Ribosomal "life" springs and dies too often for ribosomes to be significant past our stable maturation. Ribosomal life also doesn't necessary mean they are part of the same organism. Many great scientists fear connecting with other solar cells or aliens, but one should consider the nature of the solar organism. It very well is possible for a neighboring cell to destroy our cell in many ways, but this wouldn't change with having receptor contact with them.

Our role in the solar cell is in fact to mature our cell. Let us start research now. Problems arise from certain subsystems, life expectancies arise from the system as a whole. One major problem we have is a lack of microtubule system center organizing. An immature cell has a very weak microtubule system that basically provides shape for the cell. In a eukaryotic cell, the microtubule tracks are like the rail system. Motor proteins "Run" down the tube with feet like attachments carrying materials. This is more efficient than a rocket-like protein.

We can construct a huge distribution chain that in function is the same shape as eukaryotic microtubules. Carbon nanotubes are strong enough to hold their own weight for over 6000Km so this makes them great candidates for the structure. The structure will start off connecting earth to space.

A satellite will have a carbon nanotube seed in space. This satellite will be exposed to an orbit full of hydrogen ions which help the carbon nanotube to grow for currently unknown reasons though verified in the lab. The carbon ions orbiting the van allen belts will be perfect to actually create the nanotube structure itself. The satellite in orbit will have to project a magnetic field to amplify its absorption. The tube itself could even potentially be electrical and made magnetic.

The way cells make microtubules are very similar. The cell provides the components and when they are close enough they piece together. It takes some time because the precursors to the microtubule must float around in the cytoplasm. Cells also have to attach microtubules to ribosomes and other organelles. This could suggest that our moon should be the microtubule organizing center and the evidence to support this would be it's geosynchronous orbit. It could hang from the moon and into the earths atmosphere without ever coming out of it. This would mean only a portion of the world could use this transport at a time.

This shouldn't cause distress, as soon the whole solar cell will be open to us and we can begin maturing our cell. A unified global government is inevitable. Hindering is delaying, and there is a deadline for our lifespan.

Any project that we do outside of the earth is to organize the differentiation of our solar cell. This will take a lot of work to understand, but once we complete it, we will change our cell into something special. My hypothesis is that we will differentiate into a neuron type cell. Hence our ability to think and express ideas.

This project is essentially already under process and this book is one step of the ladder. Differentiation is essentially a controlled evolution of a cell. So our evolution on earth is actually controlled differentiation of our solar cell. It's controlled because evolution on earth isn't random in the sense that it leads to random outcomes, its linear due to a tendency to conclude at similar optimal organism forms or behaviors. Behaviors such as morality are beneficial on a global scale.

Religion is crude in form but intricate in details. The form of every religion on earth is in function to stabilize believers. It's not necessarily a drug to dumb down the population, but it can be. The intricate details each different individual religion differs on is essentially a small evolved difference. The united religious ideology all religions are soon evolving to have a basis in this fractal organic theory. This theory will reveal an essence of the force that drives religion. Religion has always been directed to a supreme being we are a part of. If you trace our fractal scales to beyond the the solar cell, you notice there is a solar organism and a solar organism system. Ancient cultures used to believe in multiple gods as if they believed in only solar organisms. Modern day cultures believe in one single god as if they believed in the solar organism system. This proves that our solar organism system is god to us in the sense that it is an organism we are part of and rely on to live.

Some nature based religions believed in four fundamental forces. Earth,Water,air,fire could be equivalent to the four compounds which DNA is derived from in the solar cell. Earth is carbon, air is nitrogen, water is hydrogen, and fire is oxygen. This theory shows that a god isn't directly intervening in every day encounters as most religions insist. This theory says god is merely forcing tendencies on certain moral behaviors just as it is in an organism. God isn't a human-like figure, he is a living tower we are a beam of. Essentially this theory supports ideas Albert Einstein had about the universe. Einstein insisted the universe was pantheistic or that god was the universe.

This theory impacts religion because religion used to be based to support and stabilize the earth or even small culture areas. Now religion must evolve and stabilize our whole solar cell. Through thousands of years, we will see a better support for our entire cell compared to a selfish behavior we see now with destructed environments. Hippies were having thoughts and beliefs that supported the organism system.

The removal of violence in the society is a possible adaptation. A cell in stress is equivalent to violence. Removing stress is a difficult way to solve this problem. Violence from serial killers could be removed by a societal tool. A mainstream athletic sports game could be modeled around the feeling serial killers get when they kill. This sport could be a gathering of killers and could possibly fulfill their irrational urges. This could also help law enforcement red flag potential killers and prevent crime. The game could feature the physiological feeling only they get. The game mechanics thus would serve to filter out normal people from enjoying the sport. People would play the game when a kid to avoid the possibility of understanding the real function of the game. Another method of reducing violence could be through genetic manipulation eliminating genes contributing as well as factors in the societal environment. Sneaking gene changes through diet ,toothpaste ,air filter ,forks ,deo ,soap ,shampoo ,and anything similar in that it's original to ones society. We soon will see a new way of life reflecting benefiting the solar cell.

With this ideology, one can propel the evolution of our existence in an optimal way, no trial and errors. You could incorporate this ideology into business, technology, government, prison, entertainment, and just about anything else. Business' use techniques our cells use to improve efficiency like recycling, differentiation, and replication. Business' should realize there’s a certain size at which a business is too big, and replication will help it thrive even quicker. Governments can enforce the same way as a cell, they would mainly protect the the solar cell or group of them from virus', toxins , and steroids. They will serve to preserve the environment that people live by while letting the actual people create the interactions. There will be a point at which companies are perfectly efficient, that point we will look similar to a solar cell.

Our economy on earth will soon evolve to support

a whole solar system of development. People can maintain a business due to scarcity, if things become readily available with cheapness, our economy will change from a life career to a life expression. Communism is not like the solar cell, if a ribosome regulated itself, then it would be similar, but this isn't the case. The human economy should directly resemble that of the cell.

Chapter 4- The Universe

We have ordered the fractal scales of life's existence with great luminosity. Organization is essentially all life is. The optimal format for this always comes out due to evolution. Evolution in this case is essentially controlled because true evolution can only occur if the environment changes as a whole, and we know for life on earth to be stable the environment must be stable as well.

The cellular scale is the smallest relative scale we have seen. This scale may be small to our eyes but in our bodies, cells aren't small, they're optimal. To our view, they are chaotic and blobular, but to their function they are harmonious and vivid. Cells differ from the organism scale and organism system scale in that it is setup to be the basis of any large scale organism.

The next scale of the fractal is the solar organism. This organization of cells is the most vivid to us because we are one. Living as an organic scale can feel weird, almost like it would feel to be a gear in a car. Our society has made it easy to be distracted from the confusing ideas. Organisms survive for their selves and this supports our next scale.

The organism system is a playground for organic toys to thrive and survive. Currently, Earth is the only known organism system. It's considered living because it must stabilize itself for lower forms of life to thrive. Many process' have been discovered on how this is done. The actual output of the organism system is similar to that of a cellular organelle. This puts the cycle of organic scales on repeat.

Our Solar cell is our main organic scale. We support it from the ground up. Humans essentially control a majority of the cell since we are ribosomes making machines from our resources. The solar cell is not lonely, and like the past scales it leads to an organism, or higher scale. This organism is full of solar cells. Each having diverse differentiations. Cells don't have to have life as we know it for it to be living. There doesn't need to be "aliens" for a solar system or solar cell to be part of our solar organism.

There are always impurities in organisms. Bacteria cells out number human cells in our bodies. The solar organism will have many independent cells which are living in the space of the organism.

Solar organisms survive and become stable in a complex fashion. Different types of stars collaborate with different types of planetary systems to form stable systems. The actual mechanisms and properties are very crude to us in understanding. Our solar organism is very stable and our solar cell depends on it's stability. This process will be examined in the continuation of this book series.

The solar organism system is to be included in the next addition as well. The complexity might be the same as other scales, but our relative understanding is exponentially less on each further scale, giving the illusion of complexity. This scale is too massive to feel our impact even in humanity's glory days. The size and amount of isolation we are from it give us no power or authority.

The equation of life isn't just on earth, it's color is fractalled throughout the void.

www.ingramcontent.com/pod-product-compliance
Lightning Source LLC
Chambersburg PA
CBHW071548170526
45166CB00004B/1592
* 9 7 8 1 4 9 3 5 1 1 3 0 3 *